FIREFIGHTING
IN KENT

FIREFIGHTING IN KENT

ROGER C. MARDON
AND JOHN A. MEAKINS

The
History
Press

This book is dedicated to the men and women firefighters of Kent, especially to those who have made the ultimate sacrifice.

Front Cover: A serious fire in Tonbridge High Street at teatime on 14 July 1997 called for the attendance of twenty pumps and three aerial appliances. The buildings involved were up to four floors high and included a furniture warehouse and showroom and a snooker club. (*Kent & Sussex Courier*)

Frontispiece: A Chislehurst crew is turning out in an ex-NFS 30hp Fordson 7V pump escape with Barton front-mounted pump during the early days of Kent Fire Brigade. Chislehurst fire station was closed in August 1953 and this vehicle was sold out of service in the same year. (*Kentish Times Newspapers*)

First published 2004
This edition 2006

Reprinted in 2013 by
The History Press
The Mill, Brimscombe Port,
Stroud, Gloucestershire, GL5 2QG
www.thehistorypress.co.uk

British Library Cataloguing in Publication Data.
A catalogue record for this book is available from the British Library.

ISBN 978 0 7524 3260 1

Typesetting and origination by Tempus Publishing Limited.
Printed in Great Britain.

Contents

Acknowledgements

The authors express their appreciation to all who have freely allowed photographs to be reproduced in this publication. Wherever possible, acknowledgement has been given to the source of copyright material.

Special thanks are due to Peter Coombs, QFSM, MIFireE, Chief Executive and Chief Fire Officer of Kent Fire & Rescue Service, for allowing unrestricted access to archive material held by the service in its museum.

Bibliography

G.V. Blackstone, CBE, GM, *A History of the British Fire Service* (Routledge & Kegan Paul, London, 1957)

Harry Klopper, *To Fire Committed – The History of Fire-Fighting in Kent* (Kent Council of the Fire Services National Benevolent Fund, Maidstone, 1984)

Roger C. Mardon, *An Illustrated History of Fire Engines* (Ian Allan Publishing, Hersham, 2001)

Peter H. Pearce and Pat Masters, *Fifty Vigilant Years – A History of the Kent Fire Brigade* (Maidstone, 1998)

Kent Fire Brigade, Annual Reports of the County Fire Officer (Maidstone, 1949 – 1974)

Kent & Medway Towns Fire Authority, Best Value Performance Plan 2003/2004 (Maidstone, 2003)

Introduction

Firefighting developed in recognition of the great damage and loss of life that can be caused by an uncontrolled fire. The first organised firefighting is generally attributed to the Romans. Certainly they would have been the first influence in this country when they arrived on the shores of Kent in AD43. As well as dousing the fire with water directed from simple pumps and siphons, or thrown from buckets, they used long fire hooks and axes to create fire-breaks by pulling down buildings in the path of the fire. When the Romans left Britain in the fifth century their organisation left with them and firefighting became a very hit-and-miss affair. The fire pump was lost for more than a thousand years and even the siphon, or squirt, did not reappear until the sixteenth century.

Fire devastated towns and cities for centuries and it was not until the Great Fire of London in 1666 that greater public interest in firefighting was aroused. Even then much of the impetus came from city merchants keen to avoid financial loss, and so began fire insurance. The insurance pioneers soon realised it was not in their interests to allow buildings to burn down so they set up their own private fire brigades. However, not everywhere waited for this development: Dover had a fire engine in 1700 and nine years later Rochester had two.

Saving lives from fire was very much a secondary activity of the insurance brigades, giving rise in 1836 to the formation of the Royal Society for the Protection of Life from Fire, whose purpose was to provide wheeled escape ladders in London. When the Metropolitan Fire Brigade assumed responsibility for the London escapes thirty-one years later, the Royal Society offered its services to provincial towns and supplied escapes on the condition that they were properly maintained. Canterbury and Gravesend were among the first towns in the country to benefit from such an arrangement.

From medieval times local enactments had required citizens to provide leather buckets, hooks and ladders, and fire prevention measures had been introduced.

Many country house estates and commercial enterprises operated their own fire brigades because they were remote from the nearest town brigade or because of the inadequate public service. The provision of a fire brigade was a discretionary power exercised by cities and bigger towns but it was often regarded as an unnecessary expense in smaller communities. Only in 1938, when encouraged by the threat of war, did Parliament make it a duty of local authorities to provide fire brigades, thus creating over 1,600 fire authorities across the nation. It was these local brigades and the Auxiliary Fire Service (AFS), also formed in 1938, that valiantly coped with the consequences of the Battle of Britain and much of the Blitz, until all were absorbed by the National Fire Service (NFS) in August 1941.

The fire service was returned to local authority control in 1948, with responsibility in England and Wales being given to the 146 counties and county boroughs of the day. The county of Kent and the city and county borough of Canterbury combined to form the Kent Fire Brigade on 1 April 1948, taking over seventy-nine fire stations from the NFS. Subsequent local government reorganisations have had their effect upon the brigade, most significantly in 1965 when eight fire stations in the north of the county were transferred to the newly created Greater London area. While not affecting the cover provided by the brigade, Canterbury lost its county borough status in the 1974 reorganisation and the fire brigade became the exclusive responsibility of Kent County Council. In 1998 the structure of local government changed again and Kent combined with the new Medway Towns unitary authority for fire brigade provision. The county is now served by sixty-six fire stations. Fourteen of these are manned twenty-four hours a day by whole-time crews, nine are day-crewed by full-time firefighters who are on call from home during the night, and forty-three are staffed by part-time retained personnel who respond to calls when alerted by electronic pagers.

In 1920 Gillingham Fire Brigade, serving one of the county's largest towns, answered just twenty-two fire calls. In its first year of operation to March 1949 Kent Fire Brigade responded to 5,724 calls, probably treble the county total for 1938–39. In the year to March 2003 it responded to 26,828 incidents. Less than half of these were fires but, frighteningly, more than two-thirds of the fires attended were started deliberately. Over 4,000 calls were to other emergencies of one sort or another, including nearly 1,400 road accidents. There were over 10,000 false alarms, mostly generated by automatic fire alarm systems, but more than 10 per cent were given maliciously and therefore with criminal intent.

At the time of writing this book there are Government proposals for restructuring the fire service in this country. It remains to be seen what effect the changes will have. For the moment let us just say that on 1 October 2003 the Kent Fire Brigade was renamed Kent Fire & Rescue Service, a title which, like it or not, does better reflect the statistics given above.

Roger C. Mardon,
Canterbury

John A. Meakins,
Dartford

April 2004

one

Coastal
Towns

Chief Officer H. Wells (seated) and men of Margate Fire Brigade outside the King Street fire station in 1906. CO Wells joined the brigade in 1878 and was chief officer from 1891 until his death, at the age of fifty-three, in 1914. (Kent FRS Museum)

In November 1920 Broadstairs & St Peter's Fire Brigade acquired the new Leyland motor pump seen here with brigade personnel. From left to right, back row: A. Bates, A. Smith, F. Burrows, W. Bayford, P. Moody, F. Summers. Front row: W. May (Turncock, seconded from the Water Dept), G. Austen (Superintendent), Capt. Keith-Jones (Chief Officer), L. Wilson, W. Jarman, F. Day, W. Brown, W. May. (Kent FRS Museum)

This fire engine was built in 1922 for the Borough of Dover Fire Brigade on a second-hand American Peerless chassis, which was almost certainly an ex-military example from the First World War. The conversion was carried out by Merryweather & Sons Ltd of Greenwich and the company installed one of its Hatfield reciprocating pumps at the rear. This appliance was named 'Margaret' after the Mayoress of Dover, Margaret Barwick. (Roger C. Mardon collection)

Dover fire station was built by the borough council in 1931 and is still in use today. The line-up of fire engines at Dover in 1951 is, from left to right: 1950 Commer 21A/Whitson water tender (MKO 515), 1941 Fordson 7V pump escape (GGN 809), 1937 Leyland/Metz 104ft turntable ladder, 'Rosetta', (DKP 476) and Austin K2 salvage tender with trailer pump (GLE 630). (John A. Meakins collection)

The 3,265-ton SS *Clarita Schroder* was carrying a general cargo of cars, crated foodstuffs and machinery when a serious fire broke out in the hold while the vessel was berthed at Prince of Wales Pier, Dover, on 28 July 1960. The fire was controlled within four hours by five foam branches and nine water jets but altogether the brigade was in attendance for over nineteen hours. In front of the Commer foam tender, whose cab can be seen in the foreground, is a Commer water tender built by Hampshire Car Bodies (HCB) in 1950. (Kent Photos, Dover)

The SS *Kayseri* had broken down at sea on 7 September 1961 when fire, probably caused by spontaneous combustion, was discovered in a cargo of sunflower seeds, cotton seed and oil cake. Eleven hours later the vessel was towed into Dover Harbour and during firefighting operations at Prince of Wales Pier two explosions occurred. It is evident that the trim of the ship was severely affected and she was towed out and beached, and the hold was flooded to extinguish the fire. (Kent Photos, Dover)

On 17 November 1973 the 10,000 ton MV *Cap San Antonio*, with forty-two crew, six passengers and a cargo of dangerous chemicals aboard, called for assistance while off Dover, bound from Hamburg to South America. A tug immediately went to render assistance and a Kent Fire Brigade team was taken out by helicopter to discover a serious fire. The tug took the stricken vessel in tow but entry to Dover Harbour was not permitted owing to the ship's dangerous condition and she dropped anchor two miles off Shakespeare Cliff with six firefighting tugs in attendance. More firemen were ferried to the ship by the firefighting tug *Hibernia* and the fire, by then affecting five deck levels, was controlled after eight hours' sustained effort. Unfortunately, four crew members and two passengers lost their lives. (Kent FRS Museum)

The EPL Firecracker was a new concept in British hydraulic platforms with its telescopic upper boom, reaching to a height of 23.5m (77ft). In 1981 Kent Fire Brigade took delivery of three Firecrackers mounted on Dennis F125 chassis and this new one is on a training session at Crabble Mill, Dover. They were not satisfactory from the brigade's point of view and were withdrawn from service after a few months. (John A. Meakins)

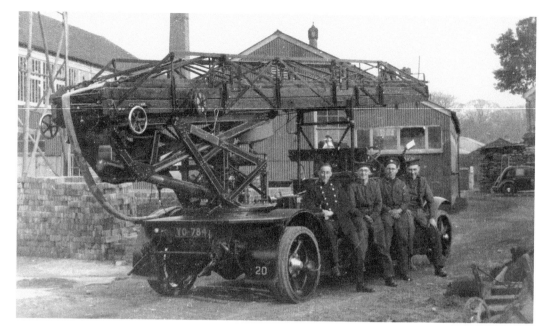

Above: This 1926 German–built Magirus 85ft turntable ladder with wooden sections, mounted on a Dennis chassis (YO 784), was purchased second-hand from London Fire Brigade by Deal Borough Council. It is seen here while stationed in Bowzell's Yard, Walmer, during the early stages of the Second World War. (Kent FRS Museum)

Left: The same appliance was being demonstrated to the public by a National Fire Service crew during the Wings for Victory week at Sandwich in 1943 when the ladder collapsed, apparently after being caught by a sudden gust of wind. The ladder had been extended to a height of 84ft with Fireman Percy Durrant Farmer at the head. Tragically, he was thrown to the ground and killed. (Kent FRS Museum)

In 1919 Folkestone Fire Brigade purchased a new Dennis pump escape, seen here on the right. This vehicle survived into the Second World War, by then running on pneumatic tyres, and subsequently languished in a scrapyard at Sissinghurst until 1980. It was then acquired for preservation and has since been restored. The vehicle on the left is a 1917 Model T Ford tender fitted out locally by Martin Walter. (Kent FRS Museum)

This breathing apparatus, being demonstrated by Folkestone Fire Brigade in October 1919, relied on a supply of air from manually-operated bellows through reinforced tubing to the smoke helmets worn by the firemen. The American GMC (General Motors Corporation) ambulance (KN 4040) was finished in green and delivered in July of the same year. Also note the hand-pushed ambulance litter. (Kent FRS Museum)

This 100ft turntable ladder was supplied new to Folkestone in 1938 and is seen here in Kent Fire Brigade livery. The all-steel Merryweather ladder was mounted on an Albion chassis and powered by a Meadows 6-cylinder petrol engine. The fire station shown was in use from the Second World War until 1988. (Kent FRS Museum)

In 1957 the ladder set was remounted by Merryweather onto this modified AEC 11.3ltr diesel-engined bus chassis. The standard power unit for this chassis was an engine of 9.6ltrs and the larger size can be recognised by the protruding radiator. Shortly after a move from Folkestone to Bromley, the vehicle was transferred to the newly enlarged London Fire Brigade in 1965 and remained in service until 1978. (Roger C. Mardon collection)

In 1905 Hythe Fire Brigade purchased this Shand Mason horse-drawn double vertical steam fire pump of 300gpm output. In the 1950s and '60s it was on show outside Medway fire station and is currently on long-term loan to a horse-drawn vehicle enthusiast in Surrey for restoration and display. (Kent FRS Museum)

Lydd Volunteer Fire Brigade was formed in 1890 and this fire station was built then to accommodate a Merryweather 22-man horse-drawn manual pump. The Commer water tender pictured was built by HCB in 1953 and was in service until 1970. The station remained in operational use until 1973 and now houses the town museum. (John A. Meakins collection)

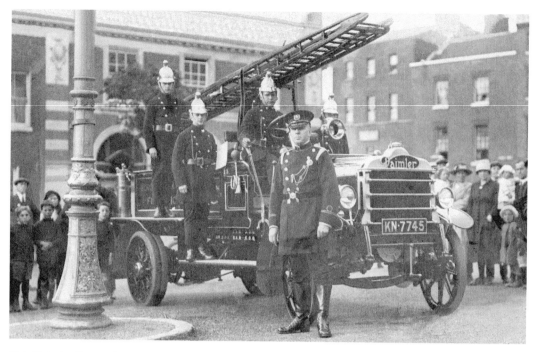

This Daimler fire tender was registered in April 1920 to Margate Fire Brigade, then under the command of CO H. Hammond. The vehicle was quite possibly a surplus War Department lorry adapted for fire brigade use. (Kent FRS Museum)

During the run up to the Second World War, Margate Fire Brigade was using its 1930 Albion/ Merryweather 85ft wooden turntable ladder (KR 3993) in a campaign to recruit AFS personnel. It is seen here in the fire station yard at King Street. The vehicle was supplied with solid tyres at a time when pneumatics were not considered to provide a sufficiently stable base for turntable ladder operation. It passed to Kent Fire Brigade from the NFS in 1948 but was promptly returned to the Home Office because of its outdated design. (Kent FRS Museum)

The Morris Ajax extension ladder from Margate's Leyland pump, depicted on the next page, is seen here pitched to a third-floor window and a charged length of delivery hose has been taken into the second floor. Margate firemen had by this time been kitted out with leather helmets and, unusually, fire tunics with collars. (Cyril Holness)

In 1936 Margate modernised its fire engine fleet by purchasing two new vehicles: this Leyland FT3 pump (DKL 363) and a Leyland FK7 escape tender (DKL 364). The crew of the pump are seen here after extinguishing a fire in the building behind. (Cyril Holness)

After the Second World War the Margate Leyland FT3 pump, as part of the NFS fleet, was rebuilt by the Home Office Workshops at Greenford as a pump escape. A Bayley wheeled escape was fitted, the rear transverse seat was removed and a hose-reel was installed, breathing apparatus lockers were provided on each side and the bodywork was extended to the outer edge of the rear wheel arches to provide more stowage. (John A. Meakins collection)

Following disposal by Kent Fire Brigade, the same vehicle was again converted, this time to a crash tender, in 1956 for Hunting Percival Aircraft Ltd at Luton Airport. Here the appliance is fitted with a blue beacon and two-tone horns in around 1965. (John A. Meakins collection)

Fire was discovered in the roof of the Cliftonville Hotel, Margate, at lunch time on 4 March 1952. Fire damage was confined almost entirely to the roof and top floor of the building. In this picture Leyland pumping appliances from Ramsgate and Margate can be seen in the foreground while Margate's Dennis/Merryweather turntable ladder (GXA 93), sheeted up with tarpaulins to prevent water from firefighting operations entering the petrol engine, is in the background. (Kent FRS Museum)

Thirteen appliances were mobilised to deal with the Cliftonville Hotel fire, which was extinguished with eight jets. Here the crew of Ramsgate's 1936 Leyland FK6 pump (DKL 467), on the right, are connecting hard suction hose to the hydrant to draw more water than would have been available from mains pressure alone. (Sunbeam Photos)

With the roof of the hotel well alight, the inevitable crowd of onlookers gathers to watch the firefighting operations. (Sunbeam Photos)

The front-line appliances at Margate in the early 1960s comprised, from left to right: 1953 Commer/Windovers water tender with 45ft alloy ladder, 1956 Hillman Husky wireless car and 1959 Bedford/Haydon/Magirus 100ft turntable ladder. A Commer/HCB water tender and Bedford AFS 'Green Goddess' emergency pump were housed across the yard in a separate building. (John A. Meakins collection)

A fire in premises at the junction of Hawley Street with Lombard Street, Margate, during the late 1970s. To the left is Margate's Dennis F43 water tender ladder, delivered new in December 1970 and sold out of service in July 1985. To the right is a 100ft Magirus turntable ladder built on a 1958 Bedford S-type chassis, which remained in service until June 1982. (Kent FRS Museum)

The building used to this day as Ramsgate fire station in Effingham Street was originally built in the eighteenth century as a dwelling and was acquired by the borough council early in the 1900s. Extensive work to convert the premises was carried out and the fire station was officially opened on 17 October 1905. The Bayley horsed escape, illustrated, and a horse–drawn manual pump remained in use until the brigade became fully motorised ten years later. (John A. Meakins collection)

Ramsgate Fire Brigade purchased a new 60hp Merryweather pump, built on a chassis with an Aster engine and running gear, in 1915. Registered KT 6801 and named 'Lord Winterstoke', this vehicle was finished in polished wood lined out in white, blue and gold. The rear–mounted Hatfield pump is a 3-cylinder reciprocating type arranged in a Y-configuration, with a copper air vessel above to even out the pulsating water flow. (Ian Scott collection)

The crew of Ramsgate's water tender ladder are dealing with a fire in a 1953 Sunbeam Talbot car in the mid-1960s. The fire appliance is a 1954 Commer 45A with HCB bodywork, a 500gpm Dennis pump and a Merryweather 45ft alloy ladder. (Kent FRS Museum)

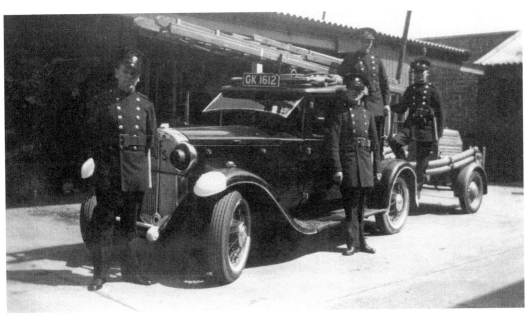

This London-registered 1930 saloon car was pressed into service as a towing vehicle, during the early years of the Second World War, with the AFS in Sheerness. (Kent FRS Museum)

In the mid-1960s Sheppey fire station was operating two Commer appliances with differing body styles. MKT 249 was a 1950 water tender bodied by James Whitson & Co. Ltd of West Drayton, and is carrying a 35ft 2-section alloy ladder. SKJ 53 was a 1953 water tender ladder bodied by HCB and is carrying a 45ft 3-section alloy ladder. The same fire station is still in use today. (John A. Meakins collection)

Sheppey fire station is not manned at night and crews are on call from home in the event of an emergency. Responding to a call in the early hours of 18 January 1973 they found their fire station and one of their fire engines badly damaged as the result of a road accident. After colliding with a milk float outside in Sheerness High Street, a 1,400-gallon milk tanker careered into the fire station, smashing through the doors and into this Dennis D-Type water tender. The driver of the milk float was killed and his assistant roundsman was seriously injured. (Kent FRS Museum)

In the evening of 26 February 1971 an electric train entering Sheerness railway station failed to stop, overrode the buffers, demolished the ticket office and came to rest on the taxi rank outside. The brigade attended with two pumps and an emergency tender with heavy lifting gear. A number of passengers were injured and a woman outside the station was crushed and fatally injured by the leading coach. (Kent FRS Museum)

The narrow-gauge Romney, Hythe & Dymchurch Railway runs regular services with one-third scale locomotives over fourteen miles of track between Hythe and Dungeness. This collision between locomotive *Black Prince* and a private car on the Burmarsh Road level crossing at Dymchurch caused derailment of the train on 7 July 2000. Firefighters released the occupant of the car from the wreckage and are seen here with paramedics rendering medical aid. (Kent FRS Museum)

Sandgate Volunteer Fire Brigade outside St Paul's church in 1884. Equipped with escape and horse-drawn manual pump, members included Firemen Hurst, Charlton, J. Sergeant (driver), H. McLachlern, T. Drayner, W. Peerless, G. Keeler (Second Officer), T. Lee, W. Fox, F. Ayres, A. Moore, W.W. Jacobs (Chief Officer), and J. Lee. (Kent FRS Museum)

Early in 1921 Sandgate formed a municipal fire brigade and the town's volunteer brigade was disbanded. After £40 had been given to the council towards the cost of equipment, the balance of the brigade funds, amounting to £232, was divided between members of the brigade according to seniority. This Crossley tender of the time would almost certainly have been converted from one of many surplus RAF vehicles that found their way on to the market after the First World War. (The Sandgate Society)

Whitstable is famous for its Native oysters and this fire engine was doubtless named 'The Native' in tribute to them. It was built on an Albion chain-drive chassis by Merryweather and taken into service by the fire brigade in 1925. (Kent FRS Museum)

The Auxiliary Fire Service was formed in 1938 in anticipation of war, and vehicles of all shapes and sizes were pressed into service as towing vehicles for the trailer pumps being issued to local authorities to supplement regular fire engines. Pictured in Cromwell Road, Whitstable, this lorry, belonging to soft drinks producer Star Minerals of Essex Street, has been given a new role to assist with the war effort. (Kent FRS Museum)

The National Fire Service operated from a station at Pavilion Mansions, Marine Parade, Tankerton, and this Fordson 7V pump escape with a Barton front-mounted pump is pictured there in the early post-war years. The premises were inherited by Kent Fire Brigade in 1948 and remained in use until the new Whitstable fire station was opened in 1969. (John A. Meakins collection)

Hurricane force winds caused disastrous flooding all along the north Kent coast on Sunday 1 February 1953 and a new sea defence wall at Whitstable proved ineffective. This is the scene at Island Wall where fire brigade trailer pumps are at work in the aftermath of the floods, to reduce water levels. Note the petrol cans against the wall for refuelling the pumps. (John A. Meakins collection)

Herne Bay has had three piers and the third, built in 1896–99, was one of the longest in the country. The Grand Pier Pavilion of 1910, accommodating a concert hall and skating rink, was severely damaged by fire while undergoing repairs and alterations on 12 June 1970. Eight pumps, an emergency tender and the brigade's mobile control unit attended the incident. (Kent FRS Museum)

This Dennis SS239/HCB-Angus water tender ladder, one of the appliances tackling a fire at premises on the corner of Oxford Street and Clifton Road, Whitstable, was new in February 1993. It was one of twenty similar machines delivered to Kent Fire Brigade in that year. (Kent FRS Museum)

The Swedish cargo vessel MV *Birkaland* reported a serious engine room fire while at sea nineteen miles off Dungeness on 7 November 1978. Under an arrangement with RAF Manston a Kent Fire Brigade team was taken out to the ship by helicopter where it was found that the fire had spread to the captain's quarters and a cargo of timber. Further personnel and equipment were flown out to the scene and after eight hours the fire was controlled. The brigade was occupied for another four hours before the ship could be towed to Dunkirk for repairs. (Chris Nelson)

two

Urban
Firefighting

Ashford Fire Engine Association assembled in 1860 at Old Trumpet Fields, two fields of about five acres in total area which straddled Beaver Road. The horse-drawn engine is a Tilley manual pump purchased in 1839. The men are wearing shako caps with 'Ashford Fire Brigade' on the cap-band. For not wearing his cap at a fire a man would have been fined 6*d*. (Kent FRS Museum)

Ashford's 1897 Merryweather Greenwich Gem steamer, turning out from the fire station at Kings Parade in the High Street. This engine is now preserved in a New Zealand fire museum. (Kent FRS Museum)

After a fire at Hothfield Common on 3 April 1926, Ashford crews are with their 1919 Daimler hose tender (KN 2054), also used for towing the steamer, and 1925 Leyland pump (KL 9478). (Kent FRS Museum)

Beckenham fire station in Bromley Road, 1920. The two Simonis appliances pictured on the next page and a 30hp Dennis pump (KT 2259) which was new in 1914 can be seen. The station building was transferred from Kent Fire Brigade to Greater London in 1965 and remained in use for another twenty years. (Chris Stone collection)

Henry Simonis & Co of Norfolk House, London WC, built a number of motor tricycle first-aid fire appliances, and Beckenham Fire Brigade took one into service in the early 1900s. The vehicle seated a driver and pillion rider and was fitted with a hose and equipment box between the front wheels. Two fire extinguishers were carried on the front of the hose-box and a folding hook ladder was carried on each side. (Kent FRS Museum)

Beckenham must have favoured Simonis products because in 1912 the urban district council, as it then was, purchased this 38-42hp Commer/Simonis pump escape (D 8019). This vehicle was fitted with a Linley patent change-speed gearbox – an early form of pre-selector transmission. (John A. Meakins collection)

Kent Fire Brigade pioneered the use of so-called light alloy ladders in Britain, in place of the then familiar wheeled escape. The original 50ft Merryweather design was soon changed for a lighter 45ft version, still dubbed the 'gut-buster', and one of the first was put into service at Beckenham in 1951. It was carried on this new Leyland Comet pump built by Windovers Ltd of Hendon. (Kent FRS Museum)

Sidcup's Leyland Comet pump, bodied by Kenex Coachworks of Ashford, is at the front of the line of appliances at a fire at Jennings, Bexleyheath, on 28 April 1960. Also just visible is a 1953 Ford Thames salvage tender of the London Salvage Corps. (John A. Meakins collection)

Bexleyheath Fire Brigade is at the scene of a factory fire in Crayford during the winter of 1902. (Kent FRS Museum)

Bromley Fire Brigade is turning out from the South Street fire station, built in 1910, with two Merryweathers supplied new at the time, an escape tender at the front and a pump behind. (Ian Scott collection)

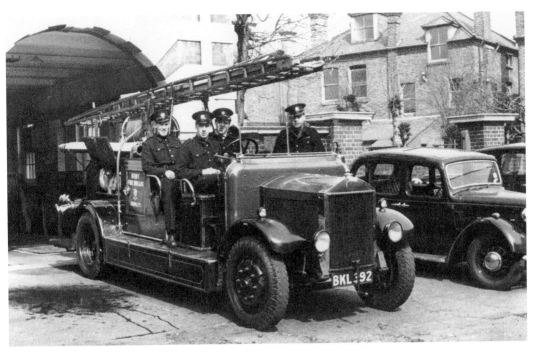

This 4-cylinder Albion/Merryweather pump was new to Bromley in 1934 and is seen here in the early days of Kent Fire Brigade. Bromley Fire Brigade stayed faithful to the Merryweather marque throughout its existence until nationalisation in 1941. (John A. Meakins collection)

This 1939 limousine-bodied Albion/ Merryweather pump escape is alongside the Bromley Magistrates' Court building. (Ian Scott collection)

Bromley's first turntable ladder was this 85ft Albion/Merryweather fitted with 550gpm pump, hose-reel and 50-gallon water tank, new in 1936. The centrifugal pump fitted to this appliance marked the brigade's departure from the reciprocating Hatfield type previously favoured. (Paul Coombe collection)

This 1943 Dennis/Merryweather 100ft turntable ladder at Bromley was inherited by Kent Fire Brigade from the NFS in 1948. The rear of the fire station had to be modified to accommodate the extra length of this vehicle over its 1936 predecessor shown in the previous picture. (John A. Meakins collection)

The anniversary of the formation of Kent Fire Brigade was marked across the county with fire station 'at home' days. Here Bromley's Dennis trailer pump receives the attention of interested boys at the 1949 open day. (John A. Meakins collection)

The Merryweather pump escape leads this 1949 turnout from Bromley fire station, followed by the pump and the turntable ladder. As the principal life-saving appliance the pump escape would always leave the station first. (John A. Meakins collection)

Fire hooks had been used since Roman times to create fire-breaks by pulling down thatch and entire buildings in the path of fire. These hooks hang in the Forrens Gate of Canterbury Cathedral. (Kent FRS Museum)

The County Fire Office came into being in 1807 and 'took into consideration the establishment of firemen' in September 1808. Officers of the County Fire Brigade at Canterbury in 1897 are, from left to right: Capt. W.G. Pidduck, First Lieutenant F. L. Henshaw, Second Lieutenant S. Truscott and Superintendent Sadleir of the Netherdale Fire Brigade, Galashiels. (Kent FRS Museum)

The Old Dover Road fire station, Canterbury, acquired by the NFS in 1943, was on the site of St Lawrence Farm near the cricket ground. The main building survives but a new fire station was opened at Upper Bridge Street in 1967. (Kent FRS Museum)

In 1949 Canterbury fire station housed Leyland FK8 pump escape, new to the city brigade in 1938, and four ex-NFS utility appliances: Fordson 7V water tender, Austin K2 salvage tender, Austin K4/Merryweather 60ft hand-operated turntable ladder and Fordson 7V hose layer. (John A. Meakins collection)

This fire at Wincheap, Canterbury, in the early 1960s was attended by a 1953 Commer/HCB water tender ladder and 1954-registered Commer/P&S Motors water tender. 'The Maiden's Head' public house survives but the Simmonds Road access to Wincheap industrial estate now occupies the site of the premises to the left. (Kent FRS Museum)

This fire destroyed a derelict warehouse in Balmoral Road, Gillingham, on 27 April 1994. Nearby houses were evacuated while fifty firemen tackled the blaze. Medway fire station's Simon Snorkel hydraulic platform is visible working from the street opposite. (*Medway News and Standard*)

A Gillingham crew is emerging from a breathing apparatus job in the early years of Kent Fire Brigade. In attendance is a Dennis 500gpm trailer pump coupled to an ex-NFS Fordson 7V mobile dam unit (GXM 263). (Kent FRS Museum)

Gillingham Fire Brigade had this Dennis Light 6 transverse-seated pump escape new in 1940. It was named 'Plewis' after J. Plewis, a former captain of the brigade, and is seen here at the Green Street fire station in 1952. Note the 1951 Fordson E83W wireless van alongside. (John A. Meakins collection)

Chatham fire station was built underneath the arches of the A2 trunk road viaduct in 1902 and still operates as a one pump retained station. In 1953 it was a divisional headquarters and the home base for Austin K2 salvage tender (GLT 494), Leyland turntable ladder (GLW 423) and Leyland FT3A pump escape (FKR 240) which was new to Bexley in 1939. Note the brigade K2 and the Austin A40 wireless van parked opposite. (John A. Meakins collection)

Chatham's Leyland TD7/Merryweather 100ft turntable ladder is in Gravesend Road, Strood. Thirty such machines were built on this half-cab bus chassis for the NFS in 1942. (John A. Meakins collection)

The 200-year-old No.2 slip of Chatham Dockyard, birthplace of HMS *Victory*, was engulfed in flame on 12 July 1966. Behind the Dockyard fire brigade Dennis F8 pump, the figurehead of Lord Nelson from the seventh HMS *Vanguard*, launched at Chatham in 1835, is rimmed in flame. Sixteen pumps attended this fire and forty-one people received minor injuries. (*Medway News and Standard*)

This was the scene at the nineteen-acre site of Dussek Bros, used for processing oils, waxes, paints and resins at Crayford, on 13 July 1964. Rivers of blazing wax ran through the site and twenty-six appliances were needed to bring the fire under control. The pump used as a control point displays a red and white chequered flag. (*Kentish Times Newspapers*)

This 60hp Dennis was new to Crayford Fire Brigade in the summer of 1921 and originally carried a Bayley 50ft wheeled escape. (Bexley Local Studies & Archive Centre)

In 1949 Crayford fire station occupied part of Timpson's coach depot and housed a 1942 Fordson 7V pump escape (GGN 721) which was sold in 1954 for £125. The station was closed by Kent Fire Brigade in September 1952. (John A. Meakins collection)

Members of Dartford Volunteer Fire Brigade, seen at Kent Road in the 1880s. From left to right: Capt. Bayliss, Second Officer Barton, Turncock Lowen, Fireman Challis, Superintendent (Constabulary?), Firemen Gray, Sandham, Barker. (John A. Meakins collection)

This fire in Dartford High Street claimed the life of an elderly lady before destroying the premises of S.F. Phillips, drapers, and badly damaging the adjoining building of Upson & Co. in 1900. This incident prompted the urban district council to improve its fire protection and a new fire station was completed in 1902. (John A. Meakins collection)

A 3-cylinder Shand Mason Equilibrium steamer, on trial at Dartford, is pumping out the Bull Hotel during flooding in 1900. The hotel is now known as the 'Royal Victoria & Bull'. (John A. Meakins collection)

Members of Dartford Fire Brigade with their new Aster/Merryweather pump in September 1912. The driver, J.A. Penfold, went on to become captain of Penge Fire Brigade and J.W. Ellingham, seated alongside, a local builder and captain of the brigade, was responsible for the construction of Bromley and Sidcup fire stations. (John A. Meakins collection)

Floods in High Street, Dartford, opposite Bullace Lane, are dealt with by the 1912 Merryweather in January 1925. Standing on the fire engine is George 'Jock' Bell who succeeded J.A. Penfold as driver. (Ivor Bell collection)

Engaged in pumping at more Dartford flooding in around 1930 is the town brigade's 1926 Leyland pump. Note the open-staired Maidstone & District bus built by Tilling-Stevens of Maidstone. (John A. Meakins collection)

Dartford's 1926 Leyland was delivered on solid tyres and is seen outside the Overy Street fire station in September 1930. In the officer's seat is Capt. William Railstone Mackney, an undertaker by trade, and in the driver's seat is George Boulding who lived at the station. (John A. Meakins collection)

The Leyland above was converted to run on pneumatic tyres in 1931 and is seen during a procession in Princes Road, Dartford. (John A. Meakins collection)

Dartford fire station in the early 1960s was the base for this 1953 Commer 45A/HCB water tender, on the left, and a reserve 1954-registered Commer 45A/P & S Motors water tender ladder, on the right. (John A. Meakins collection)

Kent's later Dennis D-type appliances were powered by the Rolls-Royce B.61 petrol engine and the last four were delivered in 1976 with the style of radiator grille seen here. OKE 907P was pictured at Dartford when in the reserve fleet. (John A. Meakins)

Erith's 1921 Leyland pump escape and 1916 Dennis pump are positioned outside the urban district council's fire station in Bexley Road. The station was totally destroyed by enemy action on 21 January 1944. (Bexley Local Studies & Archive Centre)

Confined space meant a sharp turn was required when leaving the appliance bay at Faversham, which is evident here. The old station's Leyland FK1 pump escape (AKJ 770), which was new to Chatham Fire Brigade in 1933, is shown during the 1950s. (John A. Meakins collection)

William Rose & Co. of Manchester built their first steamers in 1897 and discontinued building fire engines in 1902, but not before Gravesend Fire Brigade had bought this one. (Thames-side fire station collection)

In 1929 Gravesend Fire Brigade purchased this Albion/Merryweather MA1 fire engine with a rear-mounted Hatfield pump. It was registered KR 124 on 19 October 1929 and was based at the Town Hall fire station in the High Street. (Ian Scott collection)

This fire at 'Woolworths' store in King Street, Gravesend, attracted the attendance of twenty pumps and two turntable ladders on 25 August 1966. Fire crews worked hard in difficult circumstances to bring the fire under control and eight women staff trapped on the first floor were escorted to safety. (*Kentish Times Newspapers*)

'Woolworths' has a return frontage to the High Street where the two turntable ladders can be seen working. Both are German-built Magirus 100ft ladders; the nearest is the Commer-chassised hydraulic ladder from Gravesend (113 LKE) and the other a Bedford-chassised mechanical ladder from Medway (887 BKO). (*Kentish Times Newspapers*)

Northfleet Fire Brigade acquired this Model T Ford tender in 1923. The driver Fireman W.B. Tucker and Capt. C.W. Smith are seated. From left to right, back row: Second Officer Henry Packham, Firemen Walter Bevan, Thomas James, -?-. Front row: Firemen -?-, William Forder, ? Grey, William White, William Jackson. (Thames-side fire station collection)

A man was killed when a storage tank containing 2,000 tons of petrol exploded at the Shell Mex & BP depot, Grove Road, Northfleet, on 1 March 1965. Ten pumps, four foam tenders, a hose layer and a firefighting tug worked for four hours before the blaze was controlled. Another fourteen tanks threatened by the fire were cooled with twelve water jets and the brigade remained at the scene for five days. (*Kent Messenger*)

The Kent Fire Office was the county's first fire insurance company, founded in 1802 well over a hundred years after the first in London. The first Kent firemen were appointed in Deptford in 1802 and two engines were provided in Maidstone in 1804. This manual pump from the insurance brigade has '1883' emblazoned on the side and is seen outside the company's office in Maidstone High Street. The Kent Fire Office was taken over by Royal Insurance in 1901. (Kent FRS Museum)

Maidstone Fire Brigade bought this 65hp Dennis pump escape, registered KT 3812 and named 'Ethel', in 1914. It was fitted with a Gwynne 500gpm pump and 40-gallon water tank, and carried a 55ft Bayley escape. (Kent FRS Museum)

Above: Market Buildings provided accommodation for a fire station in Maidstone from the inception of the borough fire brigade in 1901 until 1967. A wartime Fordson 7V appliance that was converted into a water tender by the brigade workshops is shown during the early 1950s. Fireman Jack Love is looking on. (John A. Meakins collection)

Right: This 1943 Leyland TSC18 Beaver/Merryweather 100ft turntable ladder (GXA 85) demonstrated its ability as a water tower alongside the river Medway in Maidstone town centre after passing from the NFS to Kent Fire Brigade. The vehicle ended its service at Tunbridge Wells in 1961. (John A. Meakins collection)

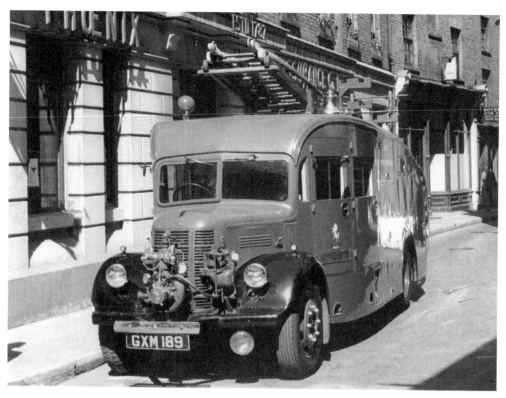

This Austin K4, outside the Phoenix Assurance Company's offices in Maidstone, was re-bodied at Home Office workshops as a limousine pump escape after the war, and delivered to Kent Fire Brigade in August 1949, valued at £3,000. It was sold, without its escape, for £62 10s in 1958. (John A. Meakins collection)

The Austin carried Kent's last operational wheeled escape, before being replaced by a Commer 45A water tender ladder, one of five ordered from P&S Motors Ltd of Teddington in 1954. The contract was assigned to Waldegrave Coach & Body Works Ltd, also of Teddington, in 1955, but the company went into liquidation before completing the contract. One machine was completed in the brigade workshops and four were eventually completed by Matador Vehicle Bodies Ltd of London W7 in 1958. (John A. Meakins collection)

Maidstone's present fire station in Loose Road was opened in 1958. From left to right: one of the Commer 45A/P & S Motors water tender ladders (SKL 823), 1952 Commer 21A/HCB pump salvage tender (OKO 550), 1943 Leyland TSC18/Merryweather turntable ladder (GXA 85), 1956 Hillman Husky wireless car (XKO 709) and 1953 Commer 7-ton/HCB hose layer (PKR 56). (John A. Meakins collection)

Maidstone Corporation Transport operated trolleybuses between 1928 and 1967. One of the routes ran past Loose Road fire station where this exercise in righting an overturned vehicle was jointly undertaken by Kent Fire Brigade and the Transport Department. (John A. Meakins collection)

The Land Rover Series I pump (WKN 567) of the Tovil Paper Mills works fire brigade can be seen at this fire involving 2,500 tons of waste paper on 11 July 1956. Appliances, including a turntable ladder, from five Kent Fire Brigade stations also attended and were engaged at the incident for five days. (Kent FRS Museum)

Maidstone's Commer VAKS/HCB-Angus K2 water tender (FKM 112D) was responding to a fire call with a crew of eight on 2 September 1970 when it skidded on the wet surface in College Road, hit a tree and burst into flames. George Stoner, a local resident later honoured for his actions, was burned as he rescued Fireman Roger Lynn, the driver, who suffered serious burns and had to have a crushed leg amputated. Fireman Malcolm Farrow sadly died from his injuries four days later. (Kent FRS Museum)

Kent's first new whole-time fire station was built in Watling Street, Chatham, at a cost of some £63,000. The site is seen here during the early stages of construction, with the houses of Star Mill Lane in the background. The station was officially opened in October 1953 and as a result existing stations at Chatham and Gillingham went over to manning by retained part-time personnel. (John A. Meakins collection)

On display after the new Medway fire station had been completed are Gillingham's Commer 21A/HCB water tender and, transferred from the station under the arches at Chatham, Leyland/Merryweather turntable ladder, Austin salvage tender and Leyland limousine pump escape. (John A. Meakins collection)

In this 1957 drill at Medway fire station, Leyland TSC18 Beaver/Merryweather 100ft turntable ladder (GXA 68), on temporary transfer from Dartford, is being used to lower a man by line from the top of the six-storey drill tower. Note the Austin K2 service van in the background. (John A. Meakins collection)

Two Bedford S/Magirus 100ft mechanically-operated turntable ladders were ordered from John Morris & Sons Ltd of Salford in 1957. This one was allocated to Medway and remained there until delivery of a new Simon Snorkel hydraulic platform in 1980. (Kent FRS Museum)

During 1969–70 eleven whole-time stations were allocated Dennis F43 appliances, powered by Rolls-Royce B.81 petrol engines of 6,522cc capacity. Medway's water tender ladder is pictured at a fire in Woodside, Wigmore in 1978. (Roger C. Mardon)

The 1916 equivalent of the Dennis F43 was this 50hp Aster/Merryweather pump escape of Rochester Fire Brigade, seen when new outside the Norman west door of the cathedral. Trade plates are displayed in the picture but the vehicle was registered KT 8271 and later named 'The Colonel'. (Roger C. Mardon collection)

Rochester Fire Brigade with a hose-reel cart and curricle escape at a devastating fire which occurred at 191 High Street on 9 March 1907. (Kent FRS Museum)

Tonbridge Fire Brigade, before the spelling was changed to avoid confusion with Tunbridge Wells, displays its manual pump, hose cart, escape and 1876 Shand Mason steamer, *c.*1898. The weather-boarded fire station was in Crown Yard at the rear of the High Street. (Kent FRS Museum)

Republic trucks were made in America between 1913 and 1929 when the Republic Motor Truck Co. Inc. merged with the commercial vehicle arm of American La France, well-known fire engine builders in the USA. Republic-badged trucks were sold in England until 1931. This Republic escape carrier (KK 5683) was new to Tonbridge Fire Brigade in 1923. (John A. Meakins collection)

This 40cwt Dennis Ace (BKL 411) was delivered to Tonbridge for use as a towing vehicle and personnel carrier in 1934. It is seen outside the fire station which was opened in 1902 on the corner of Bank Street and Castle Street. (Kent FRS Museum)

This fire station plainly exhibits its origins in the shopping frontage of High Street, Orpington, and it remained in use until the opening of a new station at Avalon Road in 1958. The ex-NFS Fordson 7V water tender (left) was sold in 1954 and the Leyland FT3A limousine pump escape (right) was new to Orpington Urban District Council (UDC) Fire Brigade as a pump in 1939. (John A. Meakins collection)

This Albion-chassised pump (GLM 292) escape at Sevenoaks was supplied to the NFS in 1942 by John Kerr & Co. (Manchester) and was inherited by Kent Fire Brigade in 1948. It was the only Albion/Kerr vehicle in the brigade and was sold in 1955. (Ian Scott collection)

Norton Court, a mansion with fifty-two rooms near Sittingbourne, was severely damaged by fire and partial collapse on 3 July 1965. On the extreme left of the picture the tail ramp of a hose layer is visible and the appliance in the centre is a Commer 45A built by Windovers Ltd in 1953 (SKE 608). (Kent FRS Museum)

This Leyland F5T2 pump (GKM 983) at Strood was new to Rochester Fire Brigade in June 1941 when it carried a wheeled escape. The fire station was replaced by a new building on the opposite side of Gravesend Road in 1965. (John A. Meakins collection)

During the Second World War all sorts of boats were converted for use by the NFS as fireboats. This vessel, apparently converted from a self-propelled barge, is set into the River Medway while in attendance at a fire in a waste paper store at Tovil, Maidstone, on 15 August 1944. (Kent FRS Museum)

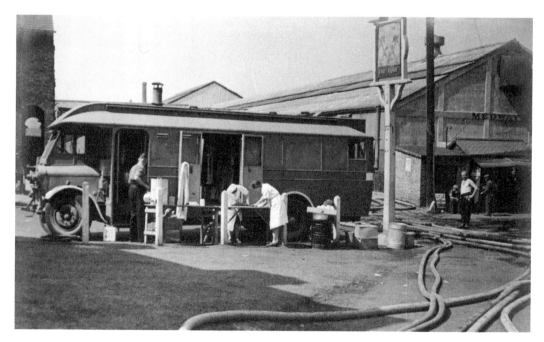

Firefighters detained for long periods at major fires need hot drinks and food to sustain them and the NFS deployed mobile kitchens and canteen vans to prepare and distribute refreshments. This mobile kitchen, converted from a single-deck bus, was set up outside 'The Rose' public house during the Tovil waste paper fire in August 1944. (Kent FRS Museum)

The only two Austin K4/Merryweather 60ft turntable ladders in the fleet (GXN 207 behind the Ford Prefect car and GXN 202 in the garage), Fordson 7V water tender (GLW 137) and, under the makeshift tarpaulin shelter, Austin K2 salvage tender (GXH 392) at Tunbridge Wells fire station, Calverley Street, in early Kent Fire Brigade days. (John A. Meakins collection)

West Wickham fire station in Glebe Way was originally completed for Beckenham Fire Brigade early in the Second World War and closed a few years after transfer from Kent to the enlarged London Fire Brigade in 1965. Here on the forecourt in the early 1960s, flanking the Commer 86A/HCB-Angus water tender ladder (761 AKT), are AFS Bedford pipe carrier (RYX 328) and Bedford S emergency pump (NYR 35). (John A. Meakins collection)

Penge Urban District Council.

FIRE BRIGADE.

Chief Officer's Report of Fire, on *30 November 1936* at *7.59 PM*

How called	*Crystal Palace Fireman by G.P.O. T. Syd 6715*
Time of Leaving	*8.0 pm* MOTOR COMBINATION
,, Return	
Place	*Crystal Palace. Crystal Palace Parade upper Norwood St.*
Description of Premises	*Crystal Palace*
Tenant	*Sole Tenants. Mecca Cafes Ltd. Baird Television Crystal Palace Trustees*
Owner	*Crystal Palace Trustees*
Estimated Loss	
Insurance	Buildings £ Contents £
	Office Insured in
Estimated Value of Property at risk (Not including adjacent buildings)	
Appliances and Men present	*PENGE. Motor Combination 7 men 1 Supt. Bear. 2 Motor pumps. 1 Motor Trailed. 11 men 1 Chief Officer CROYDON. 2 Motor Appliances. 2 Cars. 8 men 1 Chief Officer LONDON. 8. Dual purpose Machines. 58. Pumps. H. Sambrook Ladders 18. Cars 2. Fire Lorries. 7 Lorries. 2 Emergency Tenders 1 Canteen Van. 381 officers 1 men. Salvage Corps 3 Motor Tenders. 1 Motor Cal 32 men 1 Chief Officers*
How extinguished, Number of lines of Hose used, etc.	*Motor Pumps by assault. 7 Motor pumps working from Lake.*
Cause of Fire	*Un Known.*
Water Supply	
General Remarks as to Damage, etc.	*A range of buildings of one Two 1 Three floors and basement covering an area of about 1.400ft x 1.000ft used as Show rooms, Studios, Work shops, Offices 1 Stores, about two thirds of Buildings 1 Contents burned 1 fallen down, rest of Buildings 1 South Tower 1 Contents damaged by Fire Heat, Smoke 1 Water. Walls broken open. Tunnel approach slightly by Fire Heat smoke 1 Water*

Signed   Chief Officer.

While most of the appliances and men fighting the famous Crystal Palace fire on 30 November 1936 came from the London Fire Brigade, the site was within the urban district of Penge, whose brigade attended on receipt of the call at 7.59 p.m. The chief officer's report records that the superintendent and seven men attended with the motor combination appliance and that seventy-eight appliances from surrounding brigades and the London Salvage Corps also attended. (Kent FRS Museum)

three

Rural
Areas

CO J.H. Fordham and a team of Kent firefighters volunteered their services to colleagues in France when forest fires raged around Bordeaux and many lives, including those of twenty-eight firemen, were lost in 1949. The Austin K2 towing vehicle (GXH 851) and Austin 16 staff car (KKT 975) in which they travelled are seen here. (John A. Meakins collection)

This fire at Biddenden in July 1959 was attended by Headcorn's 1951 Commer/Kenex water tender (OKJ 882), seen in front of other appliances from Lenham and Hawkhurst. (John A. Meakins collection)

Motor racing enthusiast, Count Louis Zobrowski, owner of Higham Park, Bridge, donated this Rolland-Pilain touring car, re-bodied as a fire tender by Bligh Bros of Canterbury, to the village fire brigade in the 1920s. It towed a manual pump which was almost certainly a cut-down version of the Merryweather horse-drawn manual given by the Conyngham family in 1873. (Kent FRS Museum)

Count Zobrowski was killed while racing at the Monza Grand Prix in 1924 and Higham Park was bought by Walter Wigham four years later. He donated this 1911 Rolls-Royce Silver Ghost to Bridge Fire Brigade and it was kept at the fire station in Brewery Lane behind the 'Plough & Harrow' public house. It towed a Merryweather Hatfield trailer pump. (Plough & Harrow collection)

Another Silver Ghost, this time a 1921 model converted by Bonallack, served at Borough Green from 1938 and was used to tow a Dennis trailer pump. It was hidden in a barn when NFS policy dictated that it should be replaced but nevertheless it was removed from service in 1942. A third converted Rolls-Royce in the county was used as a fire tender at Cranbrook. (Kent FRS Museum)

Chislehurst Fire Brigade took delivery of a Merryweather Greenwich Gem 300pm steamer on 3 September 1910 and promptly tried it out at Prickend Pond. The arrival of a new fire engine was a major civic event in those days and a large crowd of residents had gathered round the pond to witness the display. (Bromley Libraries)

Early in the Second World War members of Hawkhurst, Cranbrook and Goudhurst fire brigades were on parade in Cranbrook High Street. Hawkhurst's 45hp Dennis pump, new in 1921, is on the left. (John A. Meakins collection)

Firemen J. Woodcock and W. Perry are included in this Cranbrook AFS crew at work in the High Street with their Dennis trailer pump, in preparation for the Second World War. (Percy Mepham collection)

A Cranbrook AFS crew is outside the Regal cinema in 1939 with a Wolseley car used to tow their trailer pump. CO Edgar Smith of the Cranbrook Rural District Council (RDC) Fire Brigade is second from the left. (Percy Mepham collection)

During the Second World War, place names were removed from fire engines, among other things, in an attempt to confuse any invading forces as to their location. The district name was painted out on this 1937 Ford V8 towing vehicle (EKJ 491) and Coventry Climax trailer pump, operated by Elham Fire Brigade. (Lyminge Historical Society)

Before it closed in April 1960, Eccles fire station housed a Commer 21A/Marsh water tender which had also served at Faversham. This appliance (MKL 746) was one of two built for the brigade by Alex J. Marsh Ltd in 1950 as part of the post-war fleet replacement programme. The other can be seen on page 111. (John A. Meakins collection)

Four pumps, a salvage tender and a hose layer attended this fire which claimed the life of an elderly woman at Oakleigh Farm, Higham, on 7 September 1953. The Commer water tender from Strood was one of four built for Kent by Folkestone-based Martin Walter Ltd in 1950. (John A. Meakins collection)

Smaller and less well-funded fire brigades often bought second-hand equipment from their better-off counterparts. The parish of Hoo St Werbergh acquired this Fiat tender from Swanscombe Fire Brigade for £10 in 1926, complete with the old Swanscombe manual pump mounted on the back. (Kent FRS Museum)

Brigades with serviceable steamers were often reluctant to incur the expense of buying a complete new motor pump. Eastry Fire Brigade was one which mounted its steam pump on a motor lorry chassis in the 1920s to benefit from the motive power of a petrol engine. (Kent FRS Museum)

Hawkhurst had a fire brigade with a four-man manual pump by 1843. This was replaced in 1882 by a larger horse-drawn engine, seen here with members of the brigade under Capt. Walter Mascall, four of whose nephews succeeded in turn to the captaincy. (Kent FRS Museum)

Members of the Longfield Parish Council brigade pose with their hose cart outside The Green Man public house in around 1910, for a series of postcards sold in aid of the National Fire Brigades Union widows and orphans fund. (John A. Meakins collection)

Many London appliances found their way to a second life elsewhere in the country and this 1911 Dennis pump was sold to Horton Kirby Parish Council in 1926. It is seen after conversion from solid tyres, under the control of Dartford RDC, in the 1930s. (John A. Meakins collection)

This Dennis Light 4 (FKR 511) was new to Dartford RDC in May 1939 and allocated to Horton Kirby to replace the appliance in the previous picture. It is shown pumping from the river Darent at Westminster Fields. (Barbara Cannell collection)

This fire started in a thirteenth-century barn at Court Lodge Farm, Lenham, and threatened ancient timber cottages and the church in the village square in the evening of 15 September 1964. It took appliances from eight stations to control the blaze. The Karrier Gamecock/HCB water tender from Hawkhurst is seen in attendance on the following day. (Kent FRS Museum)

This coach caught fire on the A20 between Lenham and Charing in the mid-1960s. Just arriving is Ashford's Commer 45A/HCB water tender ladder (RKN 902). The Three Musketeers motel is along the road behind the smoke. (Kent FRS Museum)

This wartime Dodge was elegantly refurbished, with chromed radiator surround, as a water tender for Kent Fire Brigade. It served at Cranbrook as well as Loose, where it is seen at Heath Road with a Dennis trailer pump, before being sold out of service in 1955. (John A. Meakins collection)

This Fordson hose-reel tender (GKE 384), with its pre-war Dennis trailer pump, is at Ford dealers, Haynes Bros Ltd of Maidstone, before its delivery to Maidstone RDC Fire Brigade in 1939. (Kent FRS Museum)

In its early days Kent Fire Brigade attended this fire at Addington Place with 'Vanguard', the former Maidstone borough brigade's Dennis Big 6 pump escape (FKR 821), which was new in 1939. The escape, originally carried on the 1914 Dennis, 'Ethel', has been slipped from the appliance to facilitate access to the rear-mounted pump. Also seen is a wartime Dodge mobile dam unit. (John A. Meakins collection)

Yew Tree Farm at Weavering Street, Maidstone, during the early 1950s. Once again 'Vanguard' puts in an appearance, behind the tree. (Kent FRS Museum)

Appliances from Maidstone, Borough Green and West Malling attended this fire at Nepicar Farm, Wrotham, on 24 January 1951. Two Austin A40 wireless vans are in the foreground and behind is Maidstone's Commer/Marsh water tender (MKL 749). The open pump on the right is West Malling's Dennis Light 4 (GKK 987). (Kent FRS Museum)

The Kent countryside is renowned for its hop gardens and the oast house, used for drying the hops, is a distinctive feature. A Maidstone fire crew is making up equipment following an incident at one of the numerous oasts in the area, during the early 1950s. The fire engine is a 1950 Commer 21A/Marsh water tender with a demountable Coventry Climax pump at the rear. (Kent FRS Museum)

Malling's new Fordson hose-reel tender (GKE 644) was another vehicle supplied by Haynes Bros Ltd. It is at the company's premises in Maidstone, 1939. In contrast to Maidstone RDC's Fordson from the same time, the crew would have sat facing each other in the rear body rather than on a transverse seat behind the cab. (Kent FRS Museum)

Household furniture is salvaged from an early 1950s fire at St Mary's Platt and stacked on the front lawn. Borough Green's Dennis trailer pump (left), would have been towed by the station's Commer/Whitson water tender. (Kent FRS Museum)

Prior to the arrival of the ex-London Dennis in the following picture, Seal firemen used this adapted Morris vehicle and diminutive trailer pump. (Kent FRS Museum)

This former London Fire Brigade 1915 Dennis (LH 9657) served at Seal during the early years of the Second World War under the auspices of Sevenoaks RDC. The white-edged wings aided conspicuity during the blackout. (Seal fire station collection)

Above: Southborough Fire Brigade bought this Renault-chassised pump for £600 in 1927. Until then the brigade had relied on its 1891 Shand Mason manual pump, supplemented by the Council lorry and a steamer from Tunbridge Wells. (John A. Meakins collection)

Right: Sutton-at-Hone Fire Brigade, with its headquarters in Swanley, bought a Shand Mason 200-250gpm double vertical steamer in December 1903. It was demonstrated by Mrs Truman, wife of the parish council chairman, at Birchwood Park Avenue, Swanley. Capt. A. Hewett reported, 'It worked grandly and with the coal carried in the bunkers ran on for three hours.' (Geoff Blaxall collection)

Sturry Fire Brigade gallop over the level crossing with their manual pump. (Kent FRS Museum)

Sturry Fire Brigade purchased a second-hand 260-300gpm Shand Mason steamer of 1901 vintage from Bedford and used a 1919 Hotchkiss tender to draw it. (Kent FRS Museum)

Sturry's Hotchkiss with CO Brooker aboard leads Sandwich Fire Brigade's 1919 Leyland pump in this parade, probably in Canterbury in the late 1920s or early 1930s. (Kent FRS Museum)

In the first days of Kent Fire Brigade, Sturry fire station was in Fordwich Road and fire cover was provided by an Austin K2 towing vehicle and Dennis trailer pump. (John A. Meakins collection)

Tenterden's 200gpm Shand Mason steamer, new in 1896, is outside the High Street fire station, probably just before or just after the First World War. (Kent FRS Museum)

The same fire station at Tenterden was inherited by Kent Fire Brigade in 1948 and this picture illustrates how cramped it was, even for the small Karrier Gamecock/Carmichael water tender, allocated new in 1955. A new station at Ashford Road was opened in 1971. (John A. Meakins collection)

Teynham & Lynsted's Fiat (KK 882) was originally registered in 1922 as a goods vehicle with Chas. Moakes Ltd, bakers of Gillingham, but by 1925 the body of an old horse-drawn manual pump had been mounted on the back. (Kent FRS Museum)

This timber lock-up garage was rented by the NFS to provide a fire station at Teynham during the Second World War. It was transferred to Kent Fire Brigade in 1948 and accommodated an Austin K2 towing vehicle and trailer pump. (John A. Meakins collection)

West Malling Parish Council firemen display their combination escape and hose cart, while in the station is a horse-drawn manual pump. The same building remained in use until 1967 when West Malling lost its own fire station as a result of one being opened at Larkfield. (Roger C. Mardon collection)

Used as a hearse at the West Malling funeral of Fireman H. Hooker in 1936 was this 1912 Hallford tender (D 8636), acquired by Malling RDC from Chatham Fire Brigade in 1933. (Kent FRS Museum)

West Malling's Dennis pump, registered U4 and originally with Leeds Fire Brigade, is tackling a haystack fire in the late 1930s on the site of what was to become the RAF aerodrome. Even this very early appliance, with its built-in pump, was seen as an improvement on the Hallford tender, which had been used to tow the brigade's manual engine. (Ian Ritchie collection)

A fire at The Beech Inn, Mereworth, is attended by West Malling's 1952 Commer 21A/HCB water tender in the early 1960s. (John A. Meakins collection)

Members of the early Wingham Fire Brigade in their distinctive light-coloured uniforms have extinguished a farm fire. Two of these unusual firefighting coats are preserved in the Kent Fire & Rescue Service Museum. (Kent FRS Museum)

Wingham's Bessemer tender would have been a converted United States Army First World War surplus vehicle. Bessemer was an American company which built trucks between 1911 and 1926. At the wheel is Fred Tickner, who retired after thirty-three years' service in 1952. (Kent FRS Museum)

The Wilmington section of Dartford RDC Fire Brigade originally housed their 1931 Morris Commercial pump (KJ 12) in Bentley's laundry, which is visible behind. From left to right: Harry Exeter, Dick Martin, Gilbert Bentley (Honorary Capt.), Mr Cunningham, Bob Francis (Capt.), 'Tubby' Searles, John Sutton, Mr Blackman, Harry Carpenter, Len Good. (John A. Meakins collection)

A new Fordson towing vehicle (EKP 657) with trailer pump and the horse-drawn Shand Mason manual pump which had been in service since 1909, are outside Wye fire station next to Taylor's Garage in Bridge Street, 1938. (Wye Historical Society)

Sub-Officer C. Howarth and Leading Fireman D. Foster, of Dartford retained crew, appear at an upstairs window during a six-pump fire at the Manor House country club and hotel at West Kingsdown, 29 January 1967. The roof and first floor of this fifteenth-century building were completely destroyed and half of the ground floor was severely damaged. (Kent FRS Museum)

four

Mostly
Fire Engines

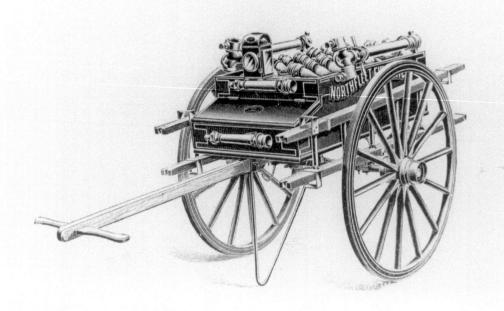

Gravesend and Northfleet are now served by Thames-side fire station in Coldharbour Road but in its early days Northfleet Fire Brigade was equipped with this Merryweather hose cart. (John A. Meakins collection)

The horse-drawn escape made its debut in 1890 and within a few years it was common for escapes to be carried on hose tenders. Such appliances were produced into the motor era and this example of Penge Fire Brigade was featured in the 1907 Merryweather catalogue. (Roger C. Mardon collection)

Bromley UDC bought this Shand Mason 260gpm double vertical steam pump at a cost of £430 in 1897. It was sold to Albert E. Reed & Co. Ltd for use at its Horton Kirby paper mill in 1919 and is now kept by Kent Fire & Rescue Service in a restored condition. (Kent FRS Museum)

Another Kent museum vehicle is the former Borough of Dartford Leyland FT4A pump escape (GKO 224) which was new in October 1939 and remained in operational use until March 1956. This appliance has an 8.84ltr petrol engine, Gwynne 700gpm pump and 40-gallon water tank. (John A. Meakins collection)

Hallford commercial vehicles were built in Dartford by J.&E. Hall Ltd from 1907 until 1925. This Hallford escape tender (D 7173), bodied by Beadles of Dartford, was registered to Foots Cray UDC in January 1912 and served at Sidcup. (Bexley Local Studies & Archive Centre)

Vulcan Motor & Engineering Co. Ltd was taken over by Tilling–Stevens of Maidstone in 1938 but this Vulcan fire tender (KK 3894), new to Sittingbourne in 1923, would have been built in Southport, Lancashire. (Kent FRS Museum)

This 1914 Dennis pump (AR 5232) was supplied to East Barnet, Hertfordshire, and was later sold to Paddock Wood Fire Brigade. The vehicle was updated locally by Ashby Commercial Vehicles for Tonbridge RDC in 1939 by the fitting of a Bedford complete front axle assembly. After the war the appliance was used by Arnolds (Branbridges) Ltd works fire brigade and is seen here in the late 1960s. (*Kent & Sussex Courier*)

Members of the NFS Overseas Contingent, which was formed to supplement the Army Fire Service for protection of military supply zones in Europe, examine the engine of a Fordson WOT3 towing vehicle at Linton training camp. (John A. Meakins collection)

A Fordson 7V mobile kitchen is providing refreshment for a large gathering of NFS personnel from 30 Fire Force. These purpose-built vehicles had coal-fired cooking stoves. (John A. Meakins collection)

All NFS personnel were encouraged to keep in trim and here a group of Kent firewomen appear to be enjoying their exercise. (John A. Meakins collection)

This 1931 Leyland Titan ex-Maidstone & District bus was converted into a mobile unit providing control, canteen and exhibition facilities by Kent Fire Brigade in 1949. The petrol-engined vehicle, with Leyland's own bodywork, continued in this role until sold in 1956. (S. M. Gillard)

Another Leyland bus, this time a fourteen-year-old Atlantean with bodywork by East Lancashire Coachbuilders, was acquired by the brigade for use as an exhibition unit in 1992. It remained in service for the next ten years spreading the fire safety message. (Roger C. Mardon)

This 1939 Bedford towing vehicle (GKE 49) was new to Strood RDC Fire Brigade and was converted to a canteen van after transfer to Kent in 1948. It is seen in its later guise, painted dark green, outside Brigade Headquarters at The Godlands, Tovil. (John A. Meakins collection)

By the 1960s the canteen van had given way to a catering box, seen here providing a welcome cup of tea from gas-heated urns. Nowadays an electric kettle and limited pre-packed refreshments are carried on most fire engines and food requirements on a larger scale at protracted incidents are bought in from local suppliers. (Kent FRS Museum)

In 1948 front-line fire engines were not equipped with radio and Kent Fire Brigade ordered ten Austin A40 Utilecon vans from Martin Walter Ltd of Folkestone for use as wireless vehicles. This one is shown as new on the forecourt at Dartford fire station. (John A. Meakins collection)

These vehicles, powered by an Austin 4-cylinder 1,200cc engine and designed to carry a load of 10cwt, were normally driven by an officer. They carried only a limited amount of firefighting equipment and a Salvus breathing apparatus set. (John A. Meakins collection)

The Bedford QL breakdown lorry is recovering a QL foam tender which was involved in an accident at Gravesend Road, Chalk, while responding to a fire call on 1 April 1951. (John A. Meakins collection)

This Leyland Beaver breakdown lorry was converted from Merryweather turntable ladder GXA 68 in 1961 and remained in this role for almost ten years. At the end of its fire brigade service it went to the gliding club at Challock where it was extensively modified for use as a winch vehicle to launch gliders. (John A. Meakins)

In the early days of Kent Fire Brigade this ex-NFS Fordson 7V (GXM 213) was rebuilt in the brigade workshops to create this water tender with a single rear-mounted hose-reel. Current thinking would require a more accessible crew door. (John A. Meakins collection)

The Kent Fire Brigade set up its own workshops in Pattenden Lane, Marden, after moving from Lyminge in the early 1950s. One of the 1957 Bedford/Magirus turntable ladders and two Dennis D-type water tenders are in for maintenance work in around 1980. (Kent FRS Museum)

Soon after its inauguration Kent Fire Brigade recognised the need to replace its vehicle fleet and in 1949 authorised the purchase of ten Commer 21A chassis, powered by the Rootes under-floor 6-cylinder petrol engine of 4,752cc. The vehicles were completed in 1950 by different bodybuilders and this is one of four by Martin Walter Ltd of Folkestone. (John A. Meakins collection)

Hampshire Car Bodies Ltd of Totton was appointed to build two of the early Commer appliances. The whole batch had a demountable fire pump at the rear which was kept connected to the 400-gallon water tank in readiness for use. (HCB-Angus Archives)

One of two Commers from the 1950 batch built by Alex J. Marsh Ltd of Ash Vale, Surrey, was on display at the County Agricultural Show, Maidstone. The mobile unit pictured on p. 105, this time with its awning in position, is visible behind. (Kent FRS Museum)

Two Commer chassis from the ten ordered in 1949 were bodied by Alfred Miles Ltd of Cheltenham, whose lightweight bodies were constructed entirely of aluminium, in contrast to the more common timber framing. (John A. Meakins collection)

A further five Commer 21A chassis were locally bodied by Kenex Coachworks of Ashford in 1951. This example (OKE 770) is carrying a 30ft Home Office wooden ladder which would have been transferred from a wartime appliance. (Kent FRS Museum)

Kenex also bodied eight Leyland Comet pumps in 1951–52, one of which was on a surplus new chassis purchased from Cambridgeshire Fire Brigade (KCE 400) and allocated to Bromley. The other Bromley Comet (OKO 664) is seen on Chislehurst Common. These vehicles had an Eaton vacuum-operated two-speed rear axle which offered intermediate ratios useful for hill-climbing. (John A. Meakins collection)

Opposite above: The transportable water unit, popularly known as the 'bikini unit', was unique to the post-war AFS and designed to pump water from sources not readily accessible to land appliances. Each vehicle carried nine portable pumps and three inflatable rafts which could be launched across soft ground or lowered into the water with the vehicle's crane. The appliance was developed in the early 1950s by the Home Office, the RFD Company and Kent Fire Brigade, the prototype being built on this Bedford S 4x2 chassis. All later appliances were built on the Commer Q4 4x4 chassis. (Kent FRS Museum)

Each raft carried three portable pumps and was powered by the reaction of a water jet from one of the pumps, giving a speed of up to six knots. This bikini raft is on the River Medway at Strood, with Rochester Castle and the cathedral across the river in the background. (John A. Meakins collection)

Motorcycles were supplied to the AFS to assist in the control of mobile fire columns of up to 144 vehicles intended to provide large-scale reinforcements anywhere in the country. They were also to be used by despatch riders for carrying messages on the fireground. This 1955 Matchless G3LS 350cc machine is at Medway fire station. (*Kent Messenger*)

Exercise 'Canopus' drew 150 auxiliary firemen from all over Kent to Chatham on 8 June 1956. A 6in pipe was laid from the old Royal Marines barracks to rough ground at the back of the town hall and up to Fort Amherst. Half a mile of plastic piping, with a pipe-bridge over Dock Road, was laid and water was on in one hour twenty four minutes. (Kent FRS Museum)

At the brigade's Linton training centre this makeshift bridge was constructed with 6in hose. The AFS vehicles beneath it are a Bedford R 4-wheel drive 'Green Goddess' emergency pump and a Land Rover Series I command car, used with a 2-wheeled trailer fitted with a field telephone cable layer. (Kent FRS Museum)

AFS crews are starting a pipe relay and a length of suction hose is about to be lowered into the stream. A metal strainer prevents debris from the watercourse being drawn into the hose and risking damage to the pump, and the strainer in turn is protected by a wicker basket. The left-hand vehicle is a Bedford QL 4x4 lorry being used as a pipe carrier. (Kent FRS Museum)

The watchroom was the nerve centre of the fire station where emergency calls were received and from which appliances were mobilised. This is the divisional watchroom at Medway in 1953 and sweeping round the console from the left are panels showing the fire situation, deployment of officers, a mobilising map showing the location of fire stations and their appliances, a panel to operate station bells and retained call-out sirens, tallies for personnel on duty and, finally, the telephone switchboard. (Kent FRS Museum)

A centralised mobilising system was brought into operation in 1966 and all emergency calls were routed to a new control room at the brigade's Maidstone headquarters. A mobilising map for the whole county occupies most of the end wall, flanked by the officer deployment and fire situation panels, and the new VF'A' scheme enabled appliances to be turned out from any fire station in Kent. The mobilising system is now computerised. (Kent FRS Museum)

The fire service employs a variety of specialist equipment according to the risks likely to be encountered. Kent and Essex fire brigades jointly brought the fireboat *Fireflair* into service to provide fire cover on the River Thames in 1961. This was a 66ft vessel built in 1957, which had undergone a refit after being with the Royal Army Service Corps for three years. Berthed at Denton Wharf, Gravesend, and manned by Kent personnel, she remained in service for five years, by which time arrangements had been made with two Thames tug companies to put fast new firefighting tugs at the disposal of the two fire brigades. (Kent FRS Museum)

The opening of the Channel Tunnel in 1994 presented another special risk to the fire brigade. Purpose-designed vehicles were built to operate in the service tunnel between the two running tunnels. Mercedes-powered units, with a driving cab at each end, accommodate interchangeable bodies for fire, ambulance, police and maintenance purposes. Four fire appliance units, built by John Dennis Coachbuilders in 1993, are stationed at each end of the tunnel. (Roger C. Mardon)

All Kent firemen were invited to submit suggestions for incorporation into the design of a new appliance in 1964. One idea was for the heel of the ladder to be positioned at a more accessible height as seen in this mock-up based on a 1952 Commer/HCB pump salvage tender at the brigade workshops. (John A. Meakins collection)

The final design provided a six-man dual-purpose appliance built by HCB-Angus on a Commer VAKS chassis. Two Proto breathing apparatus sets were stowed on a slide-out rack at a comfortable height for dressing. Suction hose and long gear were carried in a tunnel above the near-side wheel arch, accessible from the rear. This is the first machine (EKM 101C), which was allocated to Dartford. (John A. Meakins collection)

Kent Fire Brigade experimented with a light vehicle in 1971, known as a first-strike appliance, intended for rapid access to narrow and congested areas and to cope with chimney, grass and other small fires. This Ford Transit-based fire engine was equipped with a 250gpm pump and carried 105 gallons of water and a range of basic firefighting gear. It was sold in 1976 to Reed International at Aylesford paper mill. (Roger C. Mardon)

The brigade bought four of these Ford Transits, converted to 4-wheel drive, for use as accident rescue units. They were commissioned in 1985 and strategically stationed at Maidstone, Canterbury, Thames-side and Folkestone for access to the county's road network. (Roger C. Mardon)

Kent Fire Brigade's first water carrier was this 1971 Leyland Laird fuel tanker, acquired and adapted for brigade use in 1977. The vehicle was allocated to Larkfield and is seen here with the sealed dam into which its contents would be discharged for firefighting use while the tanker went to refill. It was donated in September 1991 for use as part of an aid scheme to provide domestic water supplies in Ethiopia. (John A. Meakins collection)

The latest type of water carrier, new in 2002, and now referred to as a water unit, is built on a Scania P114G-340 6x2 chassis with a 14,500ltr tank by Massey Tankers. This one is based at Ashford and an identical vehicle operates from Larkfield. (Roger C. Mardon)

Kent Fire Brigade was about to embark on a major fleet replacement programme in 1979 and purchased this white Bedford TKG/HCB-Angus CSV (crew safety vehicle) for evaluation against a Dennis RS133. The Bedford was not selected and the vehicle pictured was sold to East Sussex Fire Brigade in November 1980, and repainted red for service at Eastbourne. (Roger C. Mardon)

Before entering operational service, standard fire engines are required to undergo a tilt test to demonstrate their stability at an angle of 35 degrees, preferably 37 degrees. This MAN 12.224, bodied by John Dennis Coachbuilders Ltd, on the test bed at Chobham, Surrey in July 1999, now serves at Eastry. (Kent FRS Museum)

Special appliances in the fire service can remain inactive for significant periods and a cost-effective means of providing them has been to use a single prime mover with the ability to interchange demountable bodies, or pods. Here, a 1986 Ford Cargo prime mover is demounting an incident support unit at Strood. (Kent FRS Museum)

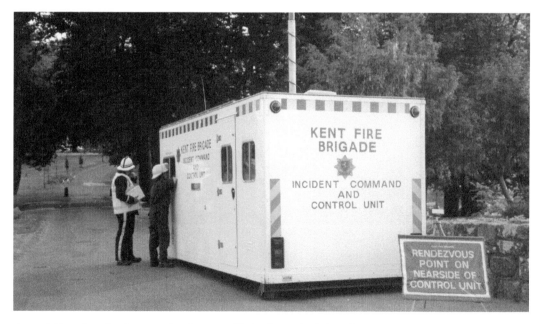

Once delivered to the fireground, a demountable body can be left at the scene to make the prime mover available for further use. Here, Strood's incident command and control unit has been set up during an exercise at Leeds Castle, near Maidstone. (Kent FRS Museum)

Another use for the demountable pod system is hose-laying. A mile of pre-connected flaked hose can be laid from the back of the vehicle as it travels at up to 20mph. The operation can be envisaged as the crew of Faversham's 1989 hose unit recover the hose after a drill. This unit has now been replaced by a purpose-built appliance with a hydraulic hose retrieval system. (Kent FRS Museum)

The Ford Cargo prime movers acquired between 1986 and 1989 have been replaced by Scania P94G-260 6x2 vehicles, with the Multilift mechanism re-chassised from the earlier Fords. This 1999 vehicle is at Ashford with one of two foam units built by John Dennis Coachbuilders in 1994. (Roger C. Mardon)

The first operational Dennis Rapier went into service with Kent Fire Brigade in 1991. This vehicle, built on a tubular stainless steel space-frame chassis and powered by an 8.3ltr Cummins diesel engine, was the result of collaboration between Dennis Specialist Vehicles and John Dennis Coachbuilders, both of Guildford. (Roger C. Mardon)

A new, roomier cab was announced by Dennis Specialist Vehicles in 1993 and over the next two years the brigade commissioned twenty-seven Mk.2 Rapiers. Those carrying rescue equipment as well as the usual range of firefighting gear were designated rescue pump ladders, such as this 1995 vehicle pictured at Tonbridge. (Roger C. Mardon)

The conventionally chassised Dennis Sabre was launched in 1995 and since then Kent has put over thirty into service, all but two being built by John Dennis Coachbuilders, like this 2001 model at Thanet. The latest deliveries make use of a larger XL cab, providing more room for the crew. (Roger C. Mardon)

Many of the county's forty-three retained fire stations, where the crew are part-time, operate lighter appliances than the whole-time stations. The latest choice of vehicle is based on a Volvo FLH 12-ton chassis with bodywork by Saxon SV of Cheshire, such as Chilham's pump which was new in 2002. (Roger C. Mardon)

The fire and rescue service is frequently called upon to rescue large animals stuck in precarious situations and Kent is now equipped with a specialist animal rescue unit. This is a four-wheel drive Mercedes-Benz Unimog U400/John Dennis Coachbuilders appliance, commissioned in 2002 and based at Faversham. (Roger C. Mardon)

Airports require a special type of appliance able to deliver large quantities of foam rapidly onto burning and crashed aircraft. This crash tender, new to London Manston Airport in 2000, is a Carmichael Cobra 2 built on a Reynolds Boughton 6-wheel drive chassis and it carries 10,500ltrs of water and 1,100ltrs of foam compound. The airport has two Cobras and an older crash tender on front-line duty. (Roger C. Mardon)

Kent Fire & Rescue Service operates three 30m (100ft) turntable ladders based at Maidstone, Canterbury and Folkestone. A reserve fourth appliance, this 1988 Scania G92/Carmichael/Magirus, was new to Canterbury where it is seen negotiating the Christchurch Gate entrance to the cathedral. (Kent FRS Museum)

Other aerial appliances stationed at Medway, Thames-side, Tunbridge Wells and Thanet are aerial ladder platforms. New vehicles, with a working height of 32m (105ft) were delivered in 2002 and 2004 on Scania 6x2 chassis with a steering rear axle and Finnish-built Bronto Skylift F32RL telescopic booms. (Roger C. Mardon)

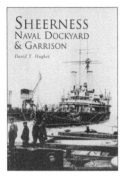

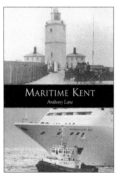